OnBoard
ACADEMICS

Traits

© 2015 OnBoard Academics, Inc
Portsmouth, NH
800-596-3175
www.onboardacademics.com
ISBN: 978-1-63096-058-2

OnBoard Academic's books are specifically designed to be used as printed workbooks or as on-screen instruction. Each page offers focused exercises and students quickly master topics with enough proficiency to move on to the next level.

OnBoard Academic's lessons are used in over 25,000 classrooms to rave reviews. Our lessons are aligned to the most recent governmental standards and are updated from time to time as standards change. Correlation documents are located on our website. Our lessons are created, edited and evaluated by educators to ensure top quality and real life success.

Interactive lessons for digital whiteboards, mobile devices, and PCs are available at www.onboardacademics.com. These interactive lessons make great additions to our books.

You can always reach us at customerservice@onboardacademics.com.

Traits and Offspring

   www.onboardacademics.com

What do babies have in common with adults?
Sort the activities.

only babies	adults and babies	only adults

sleep   breathe   crawl   work

Babies and adults have some things in common, but they also have some things that make them different from each other. Let's explore similarities and differences in other animals.

Traits and Behaviors.

A trait is a term we use to describe how an animal looks. For example this bird had black feathers so we say black feathers is a trait of this bird.

Examples of some other traits are large teeth, yellow fur, webbed feet or a long tail.

A behavior is a term we us to describe how an animal acts. This bird might eat worms so we say that eating worms is a behavior of this bird.

Examples of other behaviors might be; a tiger likes to eat deer, a dog snores when it sleeps, a duck puts its head in the water to find food, a monkey climbed tree

Remember, a trait describes what an animal might look like and a behavior describes how an animal acts.

Trait or Behavior?

Label **T** for trait or **B** for behavior.

This dog wags its tail when its happy.	The ears on this dog stand up.	This dog has brown fur.	This dog likes to bark.
This dog has four legs.	This dog has a black nose.	This dog likes to chase cats.	This dog never wags it tail when it's happy.

Lets compare an adult to an infant goose.

Study these three illustrations.

Place √ in either of the first two boxes to indicate if the description is a trait or a behavior. Then indicate if the characteristic is representative of either or both the adult and or infant with another √

	t	b	adult	infant
webbed feet				
black feathers				
flies				
swims				
black bill				
yellow feathers				

Compare the adult and infant rabbit in the same way you compared the geese.

	t	b	adult	infant
has fur				
has ears				
is blind				
drinks milk				
has four legs				
eats grass				

www.onboardacademics.com

Let's compare an adult frog with an infant frog (tadpole) in the same was as we compared the rabbits and geese.

	t	b	adult	infant
has legs				
has a tail				
has a mouth				
has big eyes				
hops				
lives under water				
swims				
eats plants				
eats insects				

Traits and Offspring Quiz

1. A trait is an identifying feature of your personal nature. True or false?

2. Which one of the following is NOT a trait passed on from parents to child.
   a. skin color
   b. hair type
   c. coordination

3. The young inherit _____ from their parents.
   a. character
   b. traits
   c. behavior

4. Which of the following is NOT a behavior?
   a. a dog chasing its tail
   b. a dog having brown fur
   c. a dog wagging its tail

5. Which of the following describes behavior?
   a. broad nose
   b. detached earlobes
   c. getting angry easily

Inherited Traits vs. Acquired Characteristics

www.onboardacademics.com

What do babies know?

√ if babies know how to do it and X if they don't.

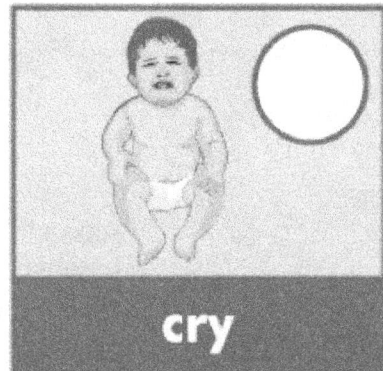

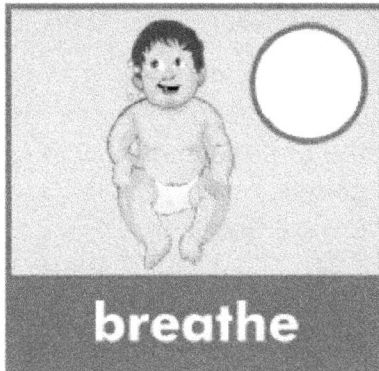

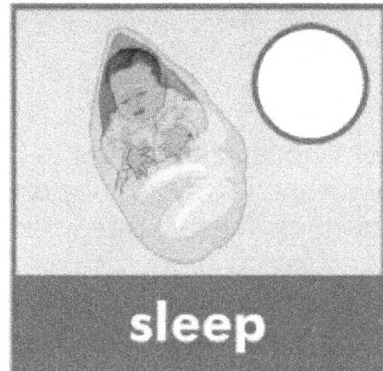

email ◯	drink ◯	cry ◯
drive ◯	breathe ◯	sleep ◯

Inherited Traits and Acquired Characteristics

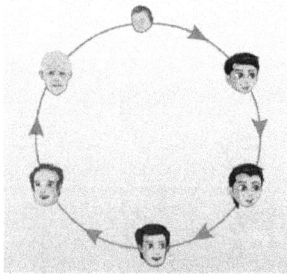

Throughout your entire life you are learning how to do new things and acquiring new skills.

For example you will learn how to walk, speak languages, drive a car, and read and write.

Along the way you will get lots of help from parents, siblings, and teachers as you learn how to do these new things and acquire new skills.

But who taught you how to breathe or sneeze or how to cry when you were a baby. No one did of course. You knew how to do these things when you were born.

We call things that you instinctively know how to do when you are born inherited traits. Things that you must learn to do are called acquired characteristics.

Acquired characteristics and inherited traits also describe a persons physical features, what you look like and your habits. You don't get to choose your eye color or you height and things like whether you have freckles, these are inherited traits.

But other things like a scar or a tattoo are acquired characteristics.

     www.onboardacademics.com

Sort these traits and characteristics.

inherited traits	acquired characteristics

Play soccer **Blink** **Smile**  **Cycle**  **Sneeze**  **Write**

Sort these physical characteristics.

> A physical characteristic describes what your body looks like. Some physical traits are inherited (you are born with them), others are acquired (you get them as you get older).

inherited traits	acquired characteristics

Pot belly   **Dimples**   **Tattoo**   **Freckles**   **Scar**   **Brown eyes**

Are these characteristics inherited or acquired over time?

√ for inherited
X for acquired over time

best friend ☐

skin color ☐

language ☐

hair texture ☐

hair length ☐

favorite food ☐

Identical Twins

Owen and Jack are identical twins.
Which traits or attributes will they
share, and which traits and attributes are likely to be different?

√ if they will be the same
X if they may be different

	Owen	Jack
hair color		
math scores		
shoe size		
height		
skating ability		
arm length		
favorite song		
eye color		
job		
number of kids		

Inherited Traits and Animals

Animals also have inherited traits. Label each animal with on of their inherited traits.

Climb

Hunt

Swim

Fly

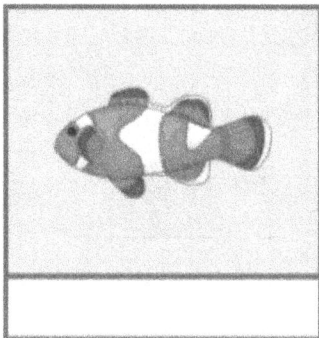

Scientists call these traits adaptations. Animals have adaptations that help them to live in their environment.

Inherited Traits vs. Acquired characteristics Quiz

1. There are two types of traits: acquitted and inherited. True or false?

2. The traits that your parents give you are called:
   a. acquired
   b. inherited

3. Which of the following could be an example of an acquired trait?
   a. scar
   b. freckles
   c. eye color

4. An example of inherited trait would be _____.
   a. dimples
   b. scar

5. Your eye color and blood type are inherited traits. True or false?

6. Hair color is an acquired trait. True or false?

7. A striped pattern on a tiger's skin is _____.
   a. acquired trait
   b. inherited trait

Adaptation

  www.onboardacademics.com

Adaptation

Traits you inherit from your parents are passed on in the form of genes. A single gene or a combination of many genes are what account for inherited traits like the shape of your nose or your eye color. Genes are the reason that you may look like other members of your family and are the reason you have your own unique characteristics.

Inherited traits often give an animal an advantage within its environment. For example the long neck of the giraffe enables it to feed at a higher level than all the other animals in its environment. Its long neck also helps it to spot predators from long distances. Traits that help an animal to survive in its environment are called adaptations. Other common examples are webbed feet, sharp teeth, acute vision or thick fur.

But, the giraffe did not always have a long neck. Fossil records suggest that a giraffe's ancestors looked like a cross between a deer and a horse. The adaptation of the giraffe that enabled it to feed more successfully in its environment occurred over a period of millions of years.

But how did this happen? Adaptations occur as a result of a process called natural selection.

To understand natural selection, lets imagine we have a family of green lizards who live in a desert habitat.

Two of the lizards, Fred and Bert, are best friends and hang out together all the time. But, Bert isn't like all the other lizards. His skin is brown and coincidentally the same color as the rocks that the lizards are relaxing on.

Unfortunately Fred and Bert's friendship comes to a sudden end when Fred was carried off by a hungry eagle. Fred was easy to spot because his green skin color stood out against the rock. Bert survived because his his skin color acted as camouflage against the rock.

Since Fred was snatched by the hungry eagle he will be unable to breed, and so in the future there will be fewer green lizards in this environment.

Since Bert was hidden from the eagle by his skin color he will be around for a longer time and more likely to breed and pass on his trait of a brown skin color to his offspring.

That's adaptation in a nutshell. Animals who inherit traits that help their survival are more likely to be around and to breed and pass along those characteristics to their future generations.

Over time the process of natural selection reenforces the continuation of characteristics that help an animal to survive and thrive in its environment until the whole population has those characteristics.

The adaptations that organisms make are directly linked to the environment in which they live. Adaptations help species to survive if their environment changes. Unlike our lizard friends, adaptations don't take place quickly but is a process that happens over many, many generations with many small changes over many millions of years.

What do you notice about the first illustration of the lizard family compared to the second?

_____

_____

Why did the eagle pick Fred instead of Bert? _____

_____

List two reasons that the giraffe adapted its neck size.

1._____

2._____

How did you parents pass traits onto you? _____

_____

Connect the animal with the described adaptation.

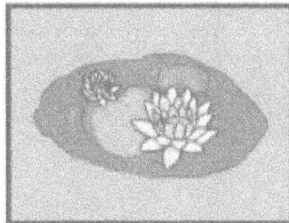

My spotted fur helps me to hide in the rain forest. I have powerful jaws for breaking turtle shells. I am short, stocky and a great swimmer.

My chiseled teeth are great for cutting down trees. My webbed feet and flat tail help me to move through rivers and ponds.

My wide flat leaves and feathery roots help me to float on the water.

My powerful legs are great for jumping and swimming. I can catch insects with my sticky tongue. I survive winter by hibernating.

How is my beak adapted to my habitat and diet?

Connect the bird with correct description.

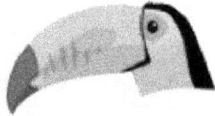

I scoop up fish with my elastic throat pouch.	My long curved beak helps me to extract nectar from flowers.

Hairy structures in my odd shaped beak help me to filter out mud and algae from the shrimp and fish that I eat.	I use my thick, chiseled like beak to extract insects and sap.

Behavior Adaptation

When an animal develops a trait to survive and thrive in its environment we call this a structural adaptation. When an organism changes what it does to survive and thrive in an environment we call this a behavioral adaptation.

For example, when you put on a coat and hat in the winter this is called a behavioral adaptation. You are adapting to colder weather.

Animals and plants also adapt their behavior to changes in their environment. This is particularly true to seasonal changes when foods become scarce. A fox will change its diet throughout the year. In spring and summer the fox eats lots of berries, grasses and insects but in the winter it eats mammals like moles, mice and squirrels.

Hibernation in another adaptation that is motivated by seasonal changes and the availability of food. Animals like chipmunks, snakes and some bears go into a deep sleep during the winter when food is no longer available and awaken in the spring when food is available again.

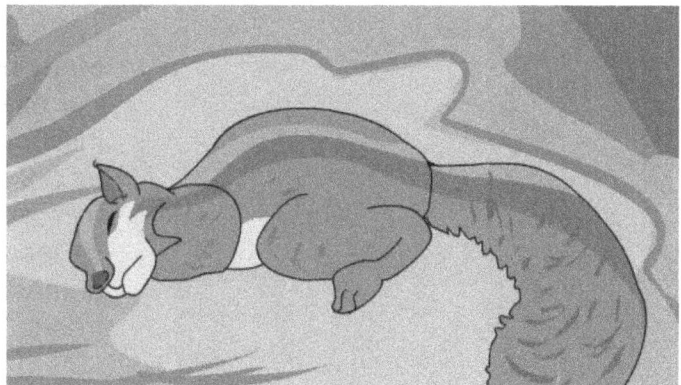

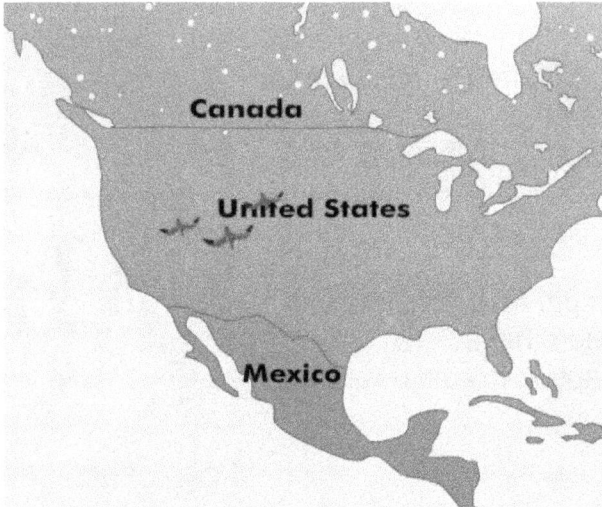

Some animals migrate, they move to another location, in order to find food and water. Canadian Geese migrate south during the winter because the ponds and other waterways that they rely on for food freeze in the winter.

Plants adapt to seasonal changes too. Some trees shed their leaves during the winter to conserve energy and retain water. Plants also adapt to their environment in other ways. Below ground, the roots of a plant will grow toward the source of water while above ground the plant will grow toward the source of sunlight. Sometimes a plant will attach itself to a wall or fence to help supporting itself away from the ground.
We call this plant behavioral adaptation tropism.

Many behavioral adaptations are instinctive. For example, when baby turtles hatch they instinctively head toward the water as soon as possible in order to avoid being picked on by predators. Other behavior adaptations can be learned. For example when we train domestic dogs, our pets, to do tricks for treats.

Identify if the adaption is structural or behavioral.

S for structural adaptation

B for behavioral adaption

1. The fur of the snowshoe hair changes from short and brown in the summer to thick, long and white in the winter. _____

2. When threatened opossums play dead, mimicking the smell and appearance of a dead animal. _____

3. Some species of squirrels bury nuts in many different areas. This is called scatter hoarding and prevents a squirrel's entire stash from being discovered by another squirrel. _____

4. A porcupine has sharp stiff quills that act as a defense agains predators. _____

5. Some harmless milk snakes look like the very poisonous coral snake. This is called mimicry and helps to deter predators. _____

6. Coyotes and badgers help each other to hunt and trap prey. Badgers do the digging and coyotes do the chasing. _____

Adaptation Quiz

1. Traits that help an animal survive in its environment are called _____.
    a. habitats
    b. adaptations
    c. characteristics
    d. genes

2. Adaptations happen over a course of a few generations. True or false?

3. Traits inherited by us from our parents are passed on in the form of _____.
    a. habitats
    b. adaptations
    c. characteristics
    d. genes

4. When an organism develops a physical trait in order to survive in an environment, we call this a _____.
    a. structural adaptation
    b. behavioral adaptation

5. A porcupine's quill is an example of a _____.
    a. structural adaptation
    b. behavioral adaptation

www.ingramcontent.com/pod-product-compliance
Lightning Source LLC
Chambersburg PA
CBHW051354200326

41521CB00014B/2575